中国水利成就系列

中国历代水利工程

水利部国际经济技术合作交流中心
中国水利水电科学研究院水利史研究所 编译

 中国水利水电出版社
www.waterpub.com.cn
·北京·

图书在版编目（ＣＩＰ）数据

中国水利成就系列. 中国历代水利工程 / 水利部国际经济技术合作交流中心, 中国水利水电科学研究院水利史研究所编译. -- 北京：中国水利水电出版社, 2020.12
ISBN 978-7-5170-8834-9

Ⅰ. ①中… Ⅱ. ①水… ②中… Ⅲ. ①水利工程—水利史—中国 Ⅳ. ①TV

中国版本图书馆CIP数据核字(2020)第169088号

审图号：GS(2020)4235号

书　　名	中国水利成就系列 **中国历代水利工程** ZHONGGUO LIDAI SHUILI GONGCHENG	
作　　者	水利部国际经济技术合作交流中心 中国水利水电科学研究院水利史研究所	编译
出版发行	中国水利水电出版社 （北京市海淀区玉渊潭南路1号D座　100038） 网址：www.waterpub.com.cn E-mail:sales@waterpub.com.cn 电话：（010）68367658（营销中心）	
经　　售	北京科水图书销售中心（零售） 电话：（010）88383994、63202643、68545874 全国各地新华书店和相关出版物销售网点	
排　　版	中国水利水电出版社微机排版中心	
印　　刷	北京博图彩色印刷有限公司	
规　　格	184mm×260mm　16开本　13.75印张（总）　191千字（总）	
版　　次	2020年12月第1版　2020年12月第1次印刷	
印　　数	0001—1500册	
总 定 价	**98.00**元（共2册）	

凡购买我社图书，如有缺页、倒页、脱页的，本社营销中心负责调换
版权所有·侵权必究

《中国历代水利工程》编委会

主　任：石秋池

副主任：金　海　朱　绛　吕　娟

委　员：谷丽雅　侯小虎　李天鹏　张林若　夏志然

鸣　谢：水利部国际合作与科技司

　　　　水利部水资源管理司

　　　　水利部农村水利水电司

　　　　中国水利水电科学研究院

如果你想了解中国，一个重要的途径就是看看从古到今中国建设的水利工程。从这些工程中，你可以看到中国的气候比如不均匀的降雨、水旱灾害发生的频率、强度……

从这里你还可以看到，生活在这里的中国人民，不畏困难、顽强生存、发展、壮大……

从天人合一的都江堰，到斗智斗勇的郑国渠，从高峡出平湖的三峡水库到从南到北跨流域的南水北调工程，从几千年到几十年、十几年，他们不仅是一座座水利工程，更是一个又一个人的传奇，一段又一段中国历史！了解了他们，你一定会感叹他们天工开物的智慧，敬佩他们为民造福的勇气！

不仅如此，或许你会更懂得中国人民，认识他们的思想，了解他们的生活方式以及他们的行为准则！

从这里，你可以走进中国，走近中国水利！更加客观地、历史地、科学地评判中国水利工程和它们对中国这片土地的重要作用！

2020 年 9 月

目录

CATALOGUE

解决中国南北水资源严重
不平衡的重要水利工程：
南水北调工程

2013年12月和2014年12月，举世瞩目的南水北调东线和中线一期工程顺利通水。长江已经成为中国中东部区域的重要水源。

中国南北区域的水资源禀赋条件差异很大。长江流域及其以南地区河川径流量占全国的80%以上，但耕地面积仅占全国的35%，北方淮河、海河和黄河流域河川径流量仅占全国的6%，但

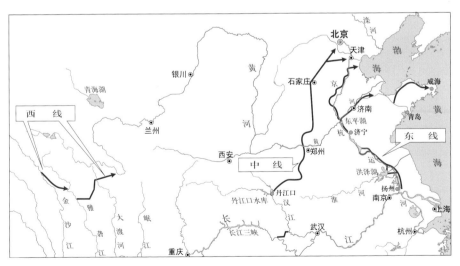

图1 南水北调工程路线图

耕地面积占全国的40%。不相匹配的水土资源组合严重影响了我国的经济发展。1952年,毛泽东主席视察黄河时提出了南水北调的构想。

经过近半个世纪的工作,在分析比较50多种规划方案的基础上,最终确定了从长江下游、中游和上游分别作为水源区,设计东、中、西三条调水路线,分别向北方不同区域调水的规划方案,通过三条调水线路,与长江、淮河、黄河、海河相互联接,构成我国中部地区水资源"四横三纵、南北调配、东西互济"的总体格局(图1)。

东线工程:利用江苏省已有的江水北调工程,逐步扩大调水规模并延长输水线路。从长江下游扬州江都抽引长江水(图2),利

图2 南水北调东线工程江都抽水站

图 3　陶岔渠首

用京杭大运河及与其平行的河道逐级提水北送，连接起调蓄作用的洪泽湖、骆马湖、南四湖、东平湖。出东平湖后分两路输水：一路向北，在位山附近经隧洞穿过黄河，输水到天津；另一路向东，通过胶东地区输水干线输水到烟台、威海。补充沿线地区的城市生活、工业和环境用水，兼顾农业、航运和其他用水。目前东线工程已经通水。东线一期工程调水主干线全长 1466.50 千米。

中线工程：从长江丹江口水库引水（图 3），沿线新建渠道，经唐白河流域西部过长江流域与淮河流域的分水岭，沿黄淮海平原西部边缘，在郑州以西穿过黄河（图 4），沿京广铁路西侧北上，基本自流到北京、天津，

向华北平原包括北京、天津在内的 19 个大中城市及 100 多个县提供生活、工业用水，兼顾生态和农业用水。目前一期工程已经实现通水。

西线工程：该工程规划在长江上游通天河、支流雅砻江和大渡河上游筑坝建库，开凿穿过长江与黄河分水岭巴颜喀拉山的输水隧洞，调长江水入黄河上游，西线工程主要是解决黄河上中游地区和渭河关中平原的缺水问题。目前西线工程还没有开始建设。

南水北调工程规划最终调水规模 448 亿立方米，其中：东线 148 亿立方米，中线 130 亿立方米，西线 170 亿立方米，建设时间需 40 ~ 50 年。

图 4 丹江口水库

图 5 南水北调中线穿黄工程

第二期

历史最悠久、仍在利用的无坝引水工程：都江堰

位于四川境内的长江一级支流——岷江中游的都江堰水利工程，始建于秦昭王末年（约公元前256—前251年），是世界水利史上的一个奇迹。因为它充分利用了当地的自然地理、水势特点，

图1　都江堰

图 2　都江堰

构筑了看似简单但却能够充分发挥其自动排沙、控制分水量和灌溉功能的水利工程，使曾经洪旱无常的四川平原变成了"水旱从人"的"天府之国"。2000年，都江堰水利工程列入《世界文化遗产名录》，2018年它又被列入《世界灌溉工程遗产名录》。

都江堰水利工程由鱼嘴、金刚堤（起分水导流作用）、飞沙堰和人字堤（起节制作用）和宝瓶口（起分水作用，见图4）等构成。最初的工程建筑材料以竹笼、木桩和卵石为主，将卵石装入竹笼内建堤堰，以木桩加固。工程在不同历史时期不断得到完善。汉代、西晋时都设有管理灌区水官。公元10世纪以来，都江堰工程形成了官方与民间在工程和用水管理上下统属且相对独立的管理体系。

得益于完善的管理体系，现代都江堰依然保留了它的基本格局，并仍在持续运用。20世纪70年代后，增加了分水枢纽，建设了水库，

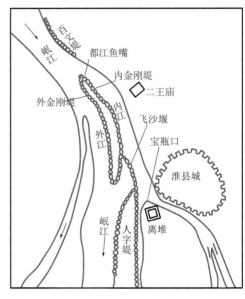

图 3 都江堰工程布置示意图

最初的直接灌溉，逐步演变为蓄引结合的灌溉方式，目前灌溉面积已经达到 1000 万亩。

都江堰水利工程已经与它所惠泽的成都平原融为一体，为纪念都江堰设计和组织建设者李冰父子而修建的二王庙等古建筑，以及分布于岷江和都江堰渠道两岸的玉垒关、索桥、观澜亭、南桥等，共同组成了都江堰文化遗产。

图 4 离堆和宝瓶口

从公元前 486 年（公元前 5 世纪）有确切记载的邗沟开凿至今，大运河已有 2500 年，由于其突出的历史、科技和文化价值，2014 年联合国教科文组织将其列入《世界文化遗产名录》。

大运河从浙江的宁波延伸至北京，历史最大长度达 2000 余千米，目前长 1750 千米。自南向北沟通了中国钱塘江、长江（太湖）、淮河、黄河和海河五个流域水系，这些水系的自然走向基本上都是从西向东。沿线最大地形高差为 50 米，连接区域的年平均降水量从 1400 毫米到 500 毫米不等。

大运河的修建最早可追溯到春秋战国时期，各诸侯国为了战争的需要，开凿了一些区间运河，分别将长江与太湖、长江与淮河、淮河与黄河、黄河与海河连通起来（图 1）。隋代开始，为了南粮北运，以上述这些连通为基础，开凿了通济渠和永济渠，系统整治了淮扬运河、江南运河，形成了以洛阳和开封（中国古代的首都）为中心的隋唐宋大运河体系（图 2）。元代定都北京，

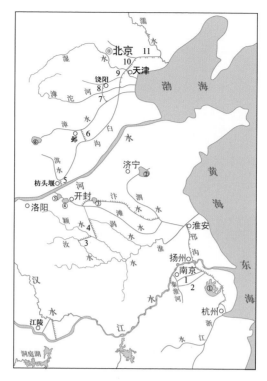

图1　6世纪时区间运河分布图

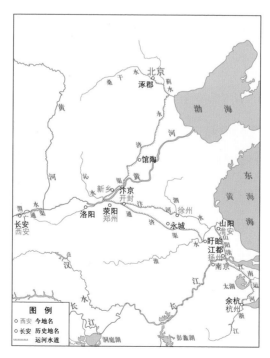

图2　隋唐宋（6—12世纪）大运河示意图

又开凿了会通河和通惠河，将隋唐宋大运河裁弯取直，形成了京杭大运河体系（图3）。

为了保证运河畅通，历代修建了很多具有地域特色的水利工程，如船闸、堤坝、水库等。这些代表了中国水利史上的最高科技水平

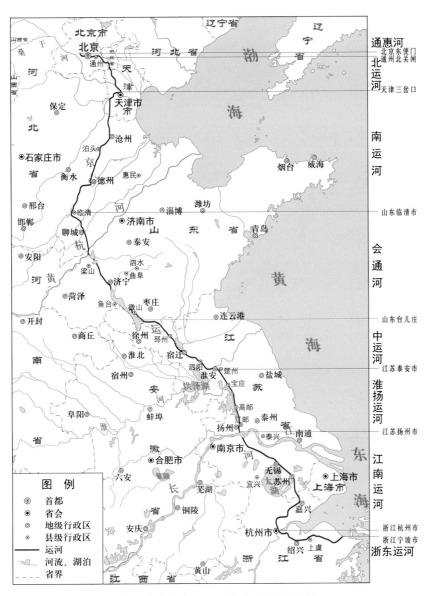

图3　元明清（12—19世纪）大运河示意图

（图4、图5）。20世纪之前，大运河不仅是交通和运粮通道，也是文化交流的重要通道。在中国的经济社会发展中发挥了重要的作用（图6）。

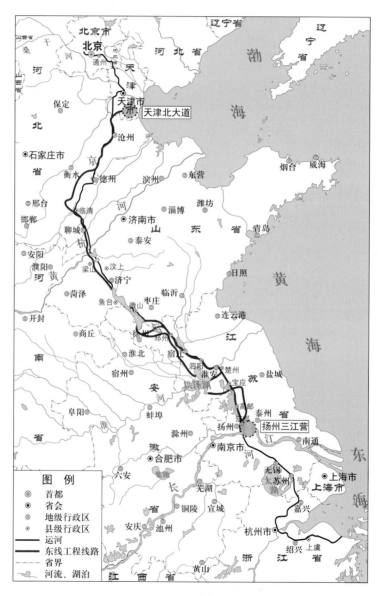

图4 大运河遗产保护

图 5　清乾隆年间英国访华使团画师所绘运河上的船只过坝图

图 6　通惠河平津闸遗址

两千年历史的海防工程：钱塘江海塘

钱塘江，浙江省最大河流，因流经古钱塘县（今杭州）而得名，以北源新安江起算，河长 588.73 千米，流经安徽省南部和浙江省，流域面积 55058 平方千米，经杭州湾注入东海。

钱塘江海塘工程是钱塘江河口两岸为抵御强潮冲刷而修筑的堤防工程。钱塘江潮被誉为"天下第一潮"，是世界一大自然奇观，

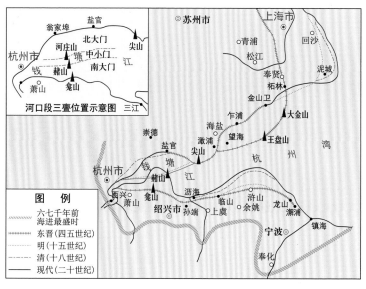

图1 钱塘江河口岸线变迁及三叠位置示意图（海宁市文物局提供）

它是天体引力和地球自转的离心作用，加上钱塘江河口的特殊地形特点，使潮流进入河口后形成涌潮。如遇天文大潮，涌潮可高达近10米，咸潮所过之地田庐尽毁。为抵御咸潮入侵，历朝历代都对海塘工程进行修缮。

钱塘江海塘修筑始于东汉（25—220年）。唐代（618—907年）钱塘江两岸已出现成规模的海塘工程，多为土塘，唐代以后渐有石塘。梁（907—923年）开平四年（公元910年），吴越王钱镠创"竹笼石塘"（图2）。北宋（960—1127年）年间出现柴塘、万柳塘、直立式石塘、斜坡式石塘等，直至明代（1368—1644年）五纵五横鱼鳞石塘建成（图3）。清代（1616—1912年）形成由基础工程、

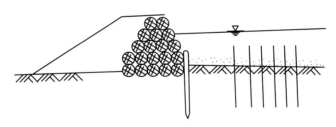

图2 吴越王钱镠"竹笼石塘"示意图

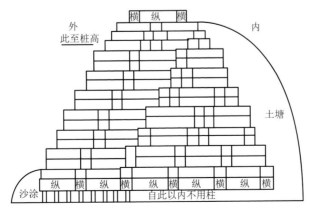

图3 明代黄光升五纵五横鱼鳞石示意图

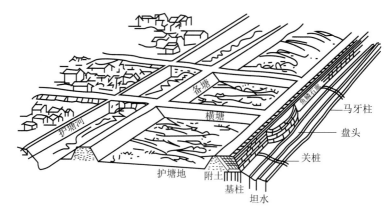

图4　清代钱塘江洪潮工程纵深防御体系示意图
（徐苏焱绘，陈方舟改绘）

主塘、备塘、邕塘（横塘）、备塘河及护塘工程组成的纵深防御体系（图4）。

清光绪三十四年（1908年）浙江海塘工程总局成立，引进西方技术，于1919—1927年，在海宁修建重力式混凝土塘（今称"洋灰塘"，图5）。中华人民共和国成立后，全面整修老海塘，又培修、加固塘外零星围堤和支堤。20世纪末开始建设标准海塘（图6、图7）。

历代政府都高度重视钱塘江海塘维护管理。自唐始就有海塘管理记述，由宋朝（960—1279年）到明朝日常维修养护均系分县自理。明代，海盐县以营造尺20丈为一号，编定石塘字号，用以明确海塘的具体位置、划段定防。至清代，设海防同知（官名，知府之佐助官），海塘开始由国家派专职人员管理。中华人民共和国成立以后，设立钱塘江管理局，两岸海塘以里程桩代替千字文碑，进一步统一管理。

图 5　钱塘江涌潮

图 6　鱼鳞大石塘

图 7　萧绍海塘

第五期

国之重器：三峡工程

坐落于长江干流上游的三峡水利水电工程于 2015 年 9 月正式竣工验收。从 2003 年 6 月下闸蓄水到现在已平安运行了 16 年，

期间经历了长江流域2010年20年一遇的大洪水和2018年2次较大的洪水，在防洪抗旱中发挥了重大作用，并输出绿色电力累计超过1000亿千瓦时。

　　人们好奇的是这样一个工程的名字为什么叫三峡。三峡是工程所在的长江这一江段的三个风景秀美的峡谷地段的名字，自西向东分别称之为瞿塘峡、巫峡、西陵峡，人们习惯将之统称为"三峡"。三峡工程建成后为长江流域增添了新的景致，毛泽东主席在其生前就描绘到："截断巫山云雨，高峡出平湖"。

　　在长江干流建设水利工程，既需要大胆设想，更需要科学求证。这一大胆的设想是孙中山先生（中国近代伟大的民主革命先行者）在1918年他的《治国方略》中首次提出的。1949年以前，民国政

图1　三峡大坝

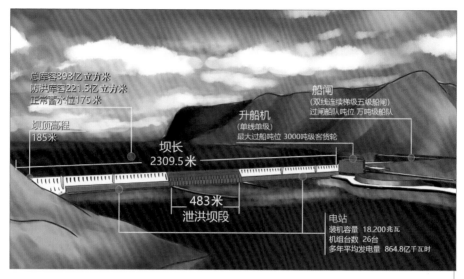

总库容393亿立方米
防洪库容221.5亿立方米
正常蓄水位175米

坝顶高程
185米

坝长
2309.5米

船闸
(双线连续梯级五级船闸)
过闸船队吨位 万吨级船队

升船机
(单线单级)
最大过船吨位 3000吨级客货轮

483米
泄洪坝段

电站
装机容量 18,200兆瓦
机组台数 26台
多年平均发电量 864.8亿千瓦时

图 2 三峡水利枢纽工程示意图

府就开始对三峡工程建设的可行性进行了研究。在 1986 年至 1992
年，中国科学家分生物、地质、泥沙、水质等 14 个专题对三峡工
程建设的可行性进行了全面的反复论证，最终确定了三峡水利工程
的防洪、发电、航运和水资源利用等主要功能和工程规模。

三峡工程主要由枢纽工程、移民工程和输变电工程三大部分组
成，坝顶高程 185 米，蓄水高程 175 米，坝长 2309.5 米，总装机
容量 2240 万千瓦，总库容 393 亿立方米，最大过航规模 3000 吨
级，静态投资 1352.66 亿元人民币。三峡工程建成后，长江荆江河
段防洪标准已提高到 100 年一遇，年均发电 848.8 亿千瓦时，相当
于减少燃煤 5000 万吨每年。

枢纽工程主要节点

1993—2016年

● 1993年1月　开始准备工程施工

● 1994年12月　枢纽工程正式开工，进入一期工程建设

● 1997年11月　大江截流，进入二期工程建设

● 2002年10月　左岸大坝全线浇筑到设计高程185米

● 2003年6月　水库蓄水至135米，首批2台机组发电，双线五级船闸试通航，进入围堰挡水发电期，三期工程建设全面展开

● 2006年5月　大坝全线浇筑到设计高程185米；10月，水库蓄水至156米水位，进入初期运行期

● 2008年10月　左、右岸电站26台机组全部投入运行，汛末开始实施正常蓄水位175米实验性蓄水

● 2012年7月　地下电站6台机组全部投产发电

● 2015年9月　长江三峡工程枢纽工程顺利通过竣工验收

● 2016年9月18日　三峡升船机正式进入试通航阶段

灵渠位于广西壮族自治区兴安县境内，是当今世界上保存最完整的古老人工运河，与都江堰、郑国渠并称为"中国秦代三大水利工程"，已于 2018 年被列入《世界灌溉工程遗产名录》。

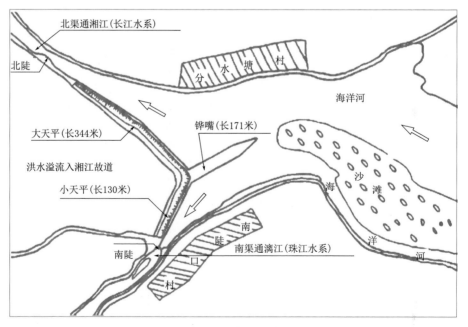

图 1　灵渠设计原理示意图

公元前219年，秦朝当政者为了统一南方区域，方便运送军粮，开工修建了一条连接海洋河—湘江（长江流域第一大支流水系）源头和大溶江—漓江（珠江流域第五大支流水系）源头的人工运河，取名灵渠。灵渠全长36.4千米，历时5年时间（公元前214年）才全部建成通航。灵渠的建成，沟通了长江流域和珠江流域，形成了中国华东华南水运网络。

灵渠主体工程包括渠首、南渠和北渠三部分，其主要设计原理是用人字坝（大、小天平坝）和铧嘴组成的分水设施将湘江截流，将其中30%的水量通过南渠引入漓江（图1）。

灵渠的科学选址、巧妙设计和精当施工，体现了我国古代水

图 2 灵渠

利工程的高超水平，特别是工程布局及水工设施系统见证了中华民族在水利工程设计方面独特的文化理解和由此形成的技术与美学传统，体现出中国古代航运水利技术的高超成就。建于渠道用来壅高水位保证通航的陡门，被世界大坝委员会专家称为"世界闸门之父"。

尽管灵渠已丧失水运功能，但灌溉功能一直延续至今，成为当地的水利命脉。

图 3 湘漓分派碑

图 4 灵渠渠首

世界灌溉工程遗产：东风堰

　　位于四川省夹江县的东风堰是一个已经持续使用350余年的以农业灌溉为主、兼有防洪等功能的综合性水利工程。2014年，因建设历史悠久、渠系配套完善、综合效益显著，入选《世界灌溉工程遗产名录》。

　　夹江，在地理上属青藏高原与四川盆地过渡段，在水文上属青衣江中游峡谷向下游平坝的过渡段。长江的支流青衣江在此穿越最后一段峡谷——千佛岩峡谷后，河道展宽，流速减缓，为夹江人民提供了肥沃的土地和丰沛的水源，加上良好的气候条件和恰当的水头落差，使夹江拥有着得天独厚的自流引水、发展灌溉农业的条件。

　　历史上，这里的百姓就仿造都江堰，在青衣江畔以竹笼装石堰水，开凿了众多的取水口，建成了纵横交错的大小渠系，使夹江与成都一样，成为"水旱从人，不知饥馑"的天府之国。但是夹江地区大多数古代堰坝多为百姓自己开凿、自己管理，不仅在总体格局上缺乏统一规划，而且管理和运行粗放，用水季节一哄而上，洪水季节各自为战，甚至以邻为壑，水事纠纷层出不穷。解决这些问题的关键是要有渠道综合规划，这个规划的实施，不仅需要政府组织，还需要一定的技术。明代正统年间（1436—1449年），时任县令陆纶成为迈出关键一步的第一人。这就是东风堰建设的缘起。

图1　东风堰灌区

明代正统年间，陆绂在青衣江左岸岔河，率领民众利用自然滩头次第以竹笼储石，无坝引水。在夹江历史上第一次将自古以来百姓自发开凿的众多堰渠归纳为两个大的渠系，变分散管理为统一管理，陆绂开凿的这两个堰渠，就是后来东风堰的两大组成部分——三大堰和八小堰。"三大堰"是灌溉县城南部的一条总干渠，在它上面有长藤结瓜似的市街堰、永通堰、龙兴堰，每个堰大约灌溉一个乡镇；"八小堰"则是灌溉县城东面、北面的一条总干渠，上面连续挡水的八个小堰。

明末清初，四川遭受惨烈战乱，夹江县城的灌溉渠系破坏严重。清康熙元年（1662年），王仕魁出任清政府任命的首位夹江县令，他重修被战争破坏的三大堰，用竹笼装卵石的办法，在上游青衣江汊流进水口修筑了一座长300余米的导水堤堰，江水壅入汊流后形成一道总堰，然后由其分水到三大堰、八小堰两

图2　水利题刻

图 3　东风堰干渠与千佛岩

个总干渠；这道东南总堰，因工程位置临近毗卢寺（今夹江县机砖厂附近），所以取名为毗卢堰。毗卢堰就是东风堰的雏形，1662 年也因此被确定为东风堰的正式开凿时间。

随着时代的发展，毗卢堰引水也渐渐满足不了灌溉需要，又先后修建了刘公堰，对三大堰、八小堰不断进行改造，上移引水口，确保了这一灌溉系统能够不断发挥作用。20 世纪 70 年代，这一灌溉系统更名为东风堰。

始建于公元前的大型无坝引水灌溉工程：郑国渠

郑国渠，大型无坝引水灌溉工程，始建于公元前246年。它的建成，对当时的秦国（公元前770—前207年）的强盛乃至秦国最终能够统一中国发挥了重要的作用。

郑国渠，位于今天的陕西省中部平原，它西引泾水东注洛水，长达300余里，建成至今已有2200多年，之后又在此基础上修建或者扩建了白渠、郑白渠、丰利渠、王御使渠、广惠渠、泾惠渠等，至今仍然发挥着灌溉功能。

泾水也就是郑国渠的引水水源——泾河，是一条含沙量很大的河流，俗称"泾水一石，其泥数斗"，实测年均输沙量2.65亿吨。当时的人们就

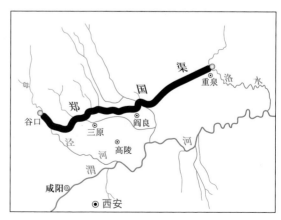

图 1 郑国渠位置

认识到，河流泥沙含有丰富的有机质，可以淤灌肥田。通过修建郑国渠，引泾河高含沙量的水，淤灌改良盐碱地。郑国渠造福了当地百姓，百姓称颂："且溉且粪，长我禾黍，衣食之师，亿万之口"。

郑国渠曾经是古代国家管理的灌溉工程。唐代中央政府颁布的水利法规《水部式》和《唐六典》，有很多专门针对郑国渠的条款，包括灌溉用水分配规则、灌溉用水优先于水磨坊运行等，已经体现出水权的意识。

郑国渠最早的引水口在泾河出峡谷的泾阳县王桥镇，200多年后引水口上移了1.3千米，称作"白渠"；又经过800多年，即公元11世纪时，渠首已经上移到泾河峡谷段；14世纪时再次上移990米，并开凿隧洞输水。1932年，重修引泾灌渠。两千多年来，泾河河床下切近20米，引泾灌渠渠首上移约5千米。

图2　郑国渠

2600余年历史的蓄水工程：芍陂

芍陂是中国古代淮河流域的水利工程，又称安丰塘，位于安徽省寿县南三十千米处，始建于公元前598年，至今仍灌溉着67万余亩（约4.47万公顷）的土地，是我国留存至今最古老的蓄水工程，比都江堰和郑国渠还要早三百多年。2015年芍陂入选《世界灌溉工程遗产名录》。

图1　芍陂1

公元前6世纪，楚国令尹孙叔敖，利用这里的低洼地势，筑堤蓄水发展灌溉。据考证，芍陂初建时蓄水面积很大，随着区域人口增加对土地的需求，芍陂水域不断被围垦成田。6世纪时蓄水面积已减少过

半；到 16 世纪末芍陂面积已萎缩至 40 多平方千米；时至今日芍陂蓄水面积仅为 34 平方千米。

芍陂最大蓄水量 9070 万立方米。历史上芍陂的水源，一部分为南面大别山发源的众多山溪水，称"山源河"；另一部分则是引淠河水，称"淠源河"。1958 年芍陂纳入淠史杭灌区之后，淠河之水通过淠东干渠进入芍陂，成为其主要水源。芍陂基本为自流灌溉。

芍陂灌溉工程体系主要由蓄水工程、灌溉水门、灌排渠系及防

图 2　芍陂 2

洪工程组成，北、东、西三面环堤，总长 26 千米，仍基本保留着 19 世纪时的工程型式和格局。

芍陂是由中央政府主持修建的大型公益工程，工程和灌溉管理则由政府和民间共同参与。历代政府均负责组织陂塘、水门和骨干渠道的修建、维护和规章制定。公元前 2 世纪的汉武帝时期，在这里设立了专门管理芍陂的陂官。20 世纪 50 年代，芍陂堤旁考古挖掘出东汉都水官铁权，见证了当时中央政府行使芍陂管理的权威。著名的水利家王景治理芍陂，制定了岁修制度，并立碑公告；灌区农民则组织管理基层灌溉用水秩序。现在芍陂的管理仍是官方和民间相结合的方式，安丰塘分局作为政府机构，负责芍陂及干支渠工程维护管理，支渠以下各级渠道及用水分配，则由受益村镇农民管理。

2600 年前芍陂的兴建成就了楚国的霸业。2600 年后的今天，芍陂灌区仍覆盖 13 个乡镇和 114 个村落，60 万人从中受益。

第十期

中国最大的堆石坝：
小浪底水利枢纽

　　小浪底水利枢纽位于河南省洛阳市以北 40 千米黄河中游最后一段峡谷的出口处，是黄河干流三门峡以下唯一能够取得较大库容的控制性工程，既可较好地控制黄河洪水，又可利用其淤沙库容拦截泥沙，进行调水调沙运用，减缓下游河床的淤积抬高。

　　小浪底水利枢纽控制流域总面积的 92.3%，来水量的 90%，来沙量的近 100%，设计目标以防洪、防凌、减淤为主，兼顾供水、灌溉、发电，蓄清排浑、兴利除害、综合利用，在黄河治理开发中具有重要的战略地位。

　　小浪底工程于 1994 年 9 月主体工程开工，1997 年 10 月截流，2000 年 1 月首台机组并网发电，2001 年年底主体工程全面完工，工程概算总投资 347.46 亿元人民币，其中利用世界银行贷款 11.09 亿美元，取得了工期提前、投资节约、质量优良的好成绩，被世界银行誉为该行与发展中国家合作项目的典范，在国际国内赢得了广泛赞誉。

　　小浪底水利枢纽由大坝、泄洪排沙建筑物、引水发电建筑物组成。大坝是斜心墙堆石坝，坝体总填筑量 5073 万立方米，是中国目前填筑量最大的堆石坝，坝高 160 米，坝顶宽 15 米，坝顶长 1667 米，正常蓄水位 275 米，库容 126.5 亿立方米，水电装机 6 台，每台 30 万千瓦，总装机容量 180 万千瓦。

小浪底水利枢纽以复杂的地形地质、特殊的水沙关系、严格的运用要求，被称为中外建设史上最具挑战性的水利工程之一。在工

图 1　小浪底水利枢纽

程建设过程中，着重加强科技攻关，推进科技创新，成功解决了一系列技术难题，取得了多项技术创新成果。

　　小浪底水利枢纽投运以来，发挥了巨大的社会效益、经济效益和生态效益，为保障黄河中下游人民生命财产安全、促进经济社会发展、保护生态与环境做出了重大贡献。

至今仍在发挥作用的古老水利工程：木兰陂

木兰陂是始建于 1064 年的水利工程，其主要目的是既挡住每日两次的咸潮，又拦蓄宝贵的淡水资源。在洪水来临时，又不影响洪水下泄，至今还在发挥作用。

木兰陂因建于木兰溪而得名。木兰溪位于中国福建中部，独流入海。建设木兰陂就是为了能够最大限度利用溪水灌溉。由于木兰陂的兴建，保证了这一区域 12 个乡镇的灌溉用水和生活用水、灌溉农田 16.3 万亩、受益人口 80 多万。

工程建设的精妙之处就在于其合适的坝高。重力闸基和挡水坝采用本地花岗石砌筑，这样的重力水工建筑物能够抵

图 1　木兰陂始建者钱四娘雕像

挡高流速的洪水冲击。闸坝基础采用木桩和抛石以保障大块砌石不发生过大的沉陷，坝堤和砌石之间用铁锭固定。900年过去了，至今木兰陂仍然保持旧时的模样。

　　这一水利工程初建者是一位名为"钱四娘"的女性，但毁于洪水。之后宋熙宁年初（1075年），由一位精通水利的高僧移至现在的位置重建。历时8年，于1083年建成。

图2　木兰陂导流堤

太湖溇港

水行于圩外　田成于圩内……

太湖位于江苏无锡、苏州和浙江湖州之间。早在西晋时期（265—316年），这里的水域面积要比现在还大，而大量人口迁移到太湖流域发现这里水多地少，特别是滩涂浅水区域面积非常大，怎样能够让水土分离并利用这部分土地，先人们想出了非常巧妙的办法，这就是人们今天称之为溇港的工程。

先人们用竹子和木头做成两道透水的墙，中间的软泥流质土被

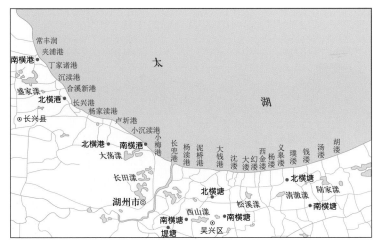

图1　太湖溇港位置

挖到挡墙的外面，而泥中的水则透过竹木围墙渗入两墙之间，再流入湖中。就这样，在太湖的周围滩涂上做成一排排这样的挡墙。久而久之，太湖周围的土地也被围成一块一块的。人们把这样的墙称之为"圩"，"水行于圩外，田成于圩内"。

两墙之间称之为溇港（河道）。为了确保安全，人们还在溇港通往太湖的交汇处修建了水闸，水闸开闭可以调节溇港水位的高低。溇港上游遇洪水时开闸将洪水入湖；溇港水位低时，引太湖水入溇港灌溉。

如果你在空中俯瞰太湖南岸溇港，可以看到所有的溇港入湖方向均为东北方，这是因为南岸区域冬天盛行西北风，人们利用它将南下泥沙冲入湖中防止溇港淤积。另外溇港大都上宽下窄，则是利用突然束窄的河道加速水流冲淤，减少溇港清淤的成本。2016 年，太湖溇港入选《世界灌溉工程遗产名录》。

图 2　太湖溇港

宁夏引黄灌区是我国四大古老灌区之一，位于黄河上游的下河沿—石嘴山两水文站之间，沿黄河两岸地形呈"J"形带状分布，至今已有两千多年的灌溉历史。宁夏平原南高北低，形成适度坡降，利用独特的地理条件，睿智的古人"天堑分流引作渠"，自此"一方擅利溉膏腴"，成就了"天下黄河富宁夏"的引黄灌溉传奇。因此，人们常说"黄河百害，唯富一套"。

早期的引黄灌区集中在银川平原南部。此后历朝历代修浚旧渠、开挖新渠，灌区范围逐渐扩大。至7世纪的盛唐时期，宁夏引黄干渠已有13条，灌溉面积达到100万亩，银川平原和卫宁平原自流灌溉体系已初步形成。11世纪，宁夏引黄灌溉在西夏王朝统治期间，工程体系和管理制度更为完备，灌溉面积扩大到160万亩。元明清时期宁夏平原干支渠道数量继续增加。至19世纪末，宁夏引黄灌溉干渠有20多条，全长1500多千米，自流灌溉面积达210万亩。

引黄古灌区工程体系包括引黄灌溉渠系、排水沟系、闸坝等控制工程。古代引黄灌溉主要采用无坝引水，渠首用长达数公里的抛石导流堤劈河引水，称作"引水湃"，湃上设有溢流坝，保障渠口防洪安全。

　　1968 年建成的青铜峡水利枢纽和 2004 年竣工的沙坡头水利枢纽，使宁夏引黄灌溉工程原有的无坝引水，全部被有坝引水工程取代，灌区渠道工程全部采用闸门节制，经过系统改造，扩展了灌溉范围、提高了灌溉保证率。目前，宁夏引黄灌区范围 12953 平方千米，总灌溉面积 828 万亩，灌区内干渠 25 条总长 2454 千米，引水能力合计 750 立方米每秒，各类控制工程 9265 座。今天，宁夏灌区已是中国 12 个商品粮基地之一。

图 1　宁夏引黄古灌区

Series of China's Achievements in Water Projects

Water Projects in China from Ancient Times

International Economic & Technical Cooperation
and Exchange Center, Ministry of Water Resources
Water History Department, China Institute of Water Resources
and Hydropower Research

中国水利水电出版社
China Water & Power Press
· BeiJing ·

The Compilation Committee
of Water Projects in China from Ancient Times

Chairwoman: Shi Qiuchi

Vice Chairmen: Jin Hai   Zhu Jiang   Lv Juan

Members: Gu Liya   Hou Xiaohu   Li Tianpeng   Zhang Linruo   Xia Zhiran

Contributing Organizations:

Department of International Cooperation, Science and Technology, Ministry of Water Resources

Department of Water Resources Management, Ministry of Water Resources

Department of Rural Water and Hydropower, Ministry of Water Resources

China Institute of Water Resources and Hydropower Research

FOREWORD

If you want to learn more about China, an effective way is to examine the country's water projects from ancient times to the present day. From these projects, you may gain a better knowledge on how the specific climate conditions in China, for example, uneven precipitation, frequent and intensive floods and droughts, have shaped the evolution of water governance in the country. Through water projects of various types and scales, you can appreciate the history of the Chinese people and civilization in tackling difficulty, surviving hardship and realizing growth and prosperity in this beautiful land.

Out of China's milestone water projects, Dujiangyan (the dam-free water diversion project) embodies the harmony between human and nature. Zhengguo Canal represents human endeavors to outwit nature. Three Gorges Reservoir creates a smooth lake among high-rising gorges. The South to North Water Transfer Project diverts water across different river basins and over a distance of one thousand kilometers. Some of these water projects boasted of a history of thousands of years, whereas others were built in modern era. However, all of them are not simply water projects by themselves. They epitomize the tremendous endeavors of legendary figures and record the splendid chapters of the Chinese history. An introduction of these water projects

will shed light on human wisdom behind these engineering wonders and help you admire human courage in defying multiple challenges and bringing benefits to the people.

Moreover, through these water projects, perhaps you will better understand the Chinese people, grasp their mindset, and appreciate their behaviors and way of life.

From these diverse water projects, you can further acquaint yourself with China and its water governance. Consequently, you will adopt a more objective, historical and scientific perspective about China's water projects and their pivotal roles in the country's past and future.

September,2020

CATALOGUE

One |

The South-to-North Water Diversion Project — A key water project overcoming the serious imbalance in water resources between North and South China

December 2013 and December 2014 witnessed the successful completion of the first phase of the Eastern and Middle Routes of the South-to-North water Transfer Project, events which attracted global attention and meant that Yangtze River has become a pivotal water source for central and eastern parts of China.

The natural conditions of water resources vary greatly between North and South China. The runoff of rivers in the Yangtze River Basin and its southern area accounts for more than 80% of the national total but it only has 35% of China's cultivated land. Meanwhile, the runoff of rivers in the Huaihe River, Haihe River and Yellow River Basins in

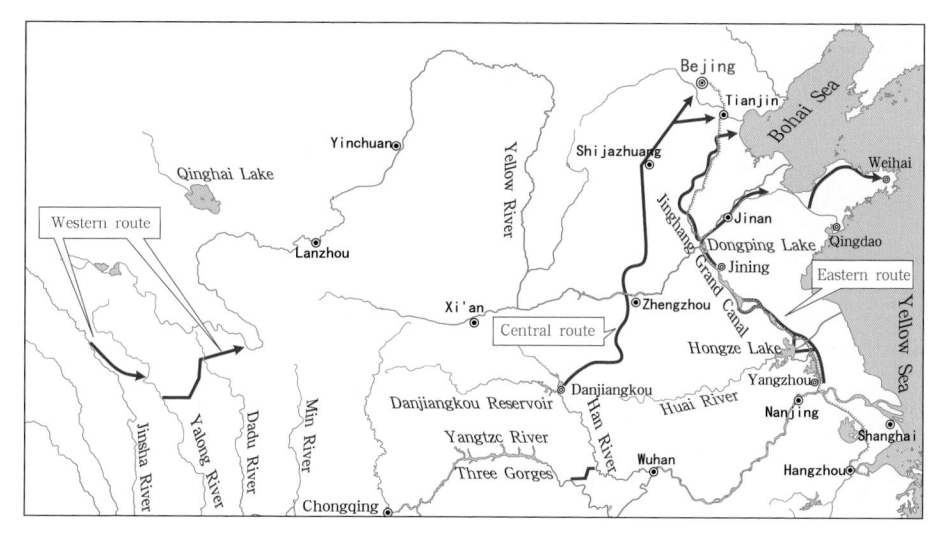

Fig.1　Roadmap of South-to-North Water Diversion Project

the North only accounts for 6% of the national total while the area has 40% of China's cultivated land. This imbalance between water and land resources seriously affects China's economic development, and led Chairman Mao Zedong to propose the concept of South-to-North Water Diversion in 1952 when he visited the Yellow River.

After nearly half a century of work, and in line with analysis and comparison on more than 50 occasions, this scheme was finalized by diverting water from the lower, middle and upper reaches of the Yangtze River to different regions in the north with eastern, middle and western diversion routes. Through these three water diversion routes, the Yangtze River, Huaihe River, Yellow River and Haihe River are inter-connected, thus constituting the overall pattern of water resource distribution in central China: four horizontal rivers and three

Fig.2　Eastern and Middle Routes of the South-to-North Water Diversion Project

Fig.3　Taocha Canal Head

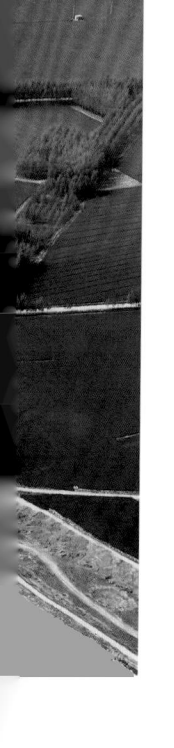

vertical canals, diversion from the south to the north, and mutual supplement between the east and the west (Fig.1).

Eastern Route: Water diversion is gradually expanded in scale and length by utilizing the existing regional water diversion project in Jiangsu Province (Fig.2). The route draws water from the Yangtze River at Jiangdu, Yangzhou, in the lower reaches of the Yangtze River, and uses the ancient Beijing-Hangzhou Grand Canal and its parallel rivers to divert water to northern China through a series of pumping stations. This route links Hongze Lake, Luoma Lake, Nansi Lake and Dongping Lake, which regulate the water supply and store water. There are two ways to transport water from Dongping Lake: to the north, passing through the Yellow River via a siphon near Weishan and

Fig.4　Danjiangkou Reservoir

terminating at Tianjin; the other is to the east, travelling through the Jiaodong water transmission line to Yantai and Weihai. The main water diversion channel and the first phase of the Eastern Route, with a length of 1,466.50 km, was completed to supplement urban, industrial and environmental water use along the route, and also takes into account agricultural, navigation and other uses.

Middle Route: Through the newly-excavated channel along the route, water is diverted from Danjiankou Reservoir (Fig.3) and passes through the watershed dividing the Yangtze and Huai River Basins at the western part of Tangbai River Basin. The route goes along the western edge of North China Plain and then crosses the Yellow River at Zhengzhou (Fig.4). The water then runs north along the western side of the Beijing-Guangzhou Railway before entering Beijing and Tianjin by gravity to meet the domestic, industrial, ecological and agricultural needs of 19 large and medium-sized cities, including Beijing and Tianjin, and more than 100 counties in the North China Plain. At present, the first phase of the Middle Route has realized water supply.

Western Route: Water is transferred from the Yangtze River to Yellow River through dams and reservoirs built on the Tongtian River, Yalong River and Dadu River, and tributaries of Yangtze River, as

well as by excavating tunnels through the Bayan Hara Mountain, the watershed between the Yangtze River and the Yellow River. This route is mainly aimed at addressing water scarcity in the upper and middle reaches of the Yellow River and the Guanzhong Plain. Construction of this route has yet to be launched.

Construction of the South-to-North Water Diversion Project will be completed in about 40-50 years. The completed project will be capable of transferring 44.8 billion m^3 of water, 14.8 billion m^3 of which from the Eastern Route, 13 billion m^3 from the Central Route and 17 billion m^3 from the Western Route.

Fig.5   Project of Crossing the Yellow River in the Middle Route of South-to North Water Diversion Project

Two

Dujiangyan — The oldest functioning dam-free water diversion project

Located in Sichuan Province in the middle reaches of the Min River, the first tributary of the Yangtze River, Dujiangyan was built at the end of the reign of King Zhao of the Qin Dynasty (about 256 B.C. to 251 B.C.). This is regarded as a miracle in the world's history of water resources which transformed the western Sichuan plain into "a land of abundance" and freed it from flood and drought, because it takes full advantage of the area's natural, geological and water characteristics.

Fig.1　Dujiangyan

Fig.2   Dujiangyan

Therefore, this seemingly simple water project can give full play to its functions in automatic sediment discharge, control of water distribution and irrigation. It was added to the UNESCO World Cultural Heritage List in 2000 and the World Irrigation Project Heritage List in 2018.

Dujiangyan is composed of the Fish Mouth and Diamond Dike (diversion and drainage function), the Flying Sand Weir and Miter Dike (control function), and Baopingkou (diversion function, Fig.4). The original construction materials were mainly bamboo cages, wooden piles and pebbles. The pebbles were loaded into the bamboo cages to build dikes and weirs, and reinforced with the wooden piles. The project has been continuously improved throughout history. During the Han Dynasty (202 B.C. to 220 A.D.) and the Western Jin Dynasty (266 A.D. to 316 A.D.), dedicated water officers were appointed in irrigation districts. Since the 10th century, a relatively independent management

system for the engineering and water use management of Dujiangyan has been formed, under the joint stewardship of the government and the people.

Thanks to its excellent management system, Dujiangyan remains operational to this day. Since the 1970s, water diversion facilities have been added and reservoirs have been built. It has evolved from providing

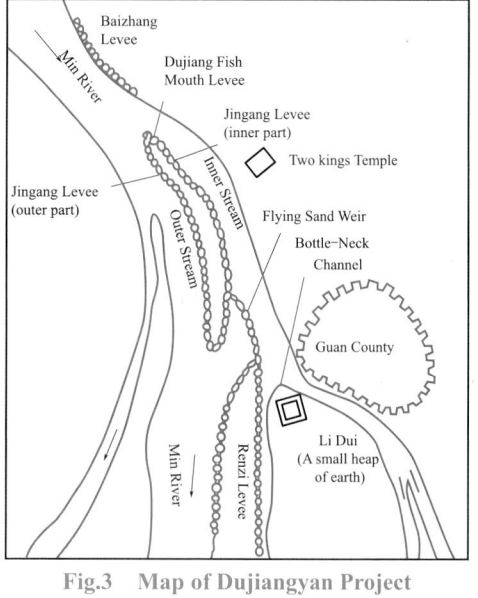

Fig.3   Map of Dujiangyan Project

direct irrigation into the combination of storage and diversion. Its current irrigation area has reached 10 million mu (667 thousand ha).

Dujiangyan has become the lifeline of the Chengdu Plain it nourishes. Ancient buildings including the Two Kings Temple built in memory of Li Bing and his son, the designer and organizer of Dujiangyan, as well as the Yulei Pass, Suspension Bridge, Guanlan Pavilion and Southern Bridge on both sides of Minjiang River and Dujiangyan Channel all constitute Dujiangyan's cultural heritage.

Fig.4   Bottle-Neck Channel

Three |

Grand Canal of China — The world's oldest and largest functioning water project

According to historical records, the Grand Canal of China has been in use for nearly 2,500 years, since the excavation of the Hangou Canal in 486 B.C. It was added to the UNESCO World Heritage List in 2014 due to its exceptional historical, technological and cultural value.

Starting at Ningbo in Zhejiang Province, the Grand Canal extends northwards for a maximum of 2,000 kilometers before terminating at

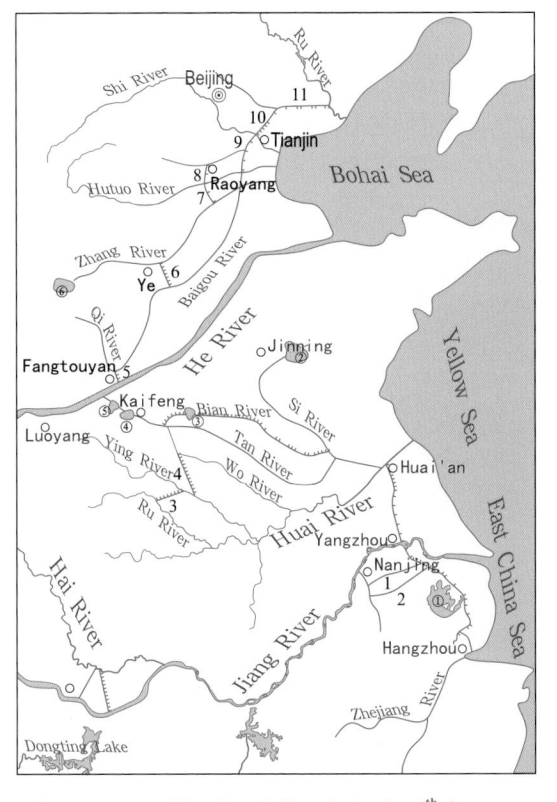

Fig.1   Map of Regional Canals in the 6th Century

Beijing. The current length of the canal is 1,750 kilometers. The Grand Canal runs across China's five river basins from the south to the north, linking the Qiantang River, the Yangtze River (Taihu Lake), the Huai River, the Yellow River and the Hai River, all of which generally flow from the west to the east. The greatest variation in altitude along the route of Grand Canal is 50 meters and the annual precipitation ranges from 500 to 1,400 millimeters.

The excavation of Grand Canal can be traced back to the Spring and Autumn and the Warring States Periods (770 B.C. to 221 B.C.). At that time, due to military needs, the warring states initiated the construction of some regional canals, connecting the Yangtze River and Taihu Lake,

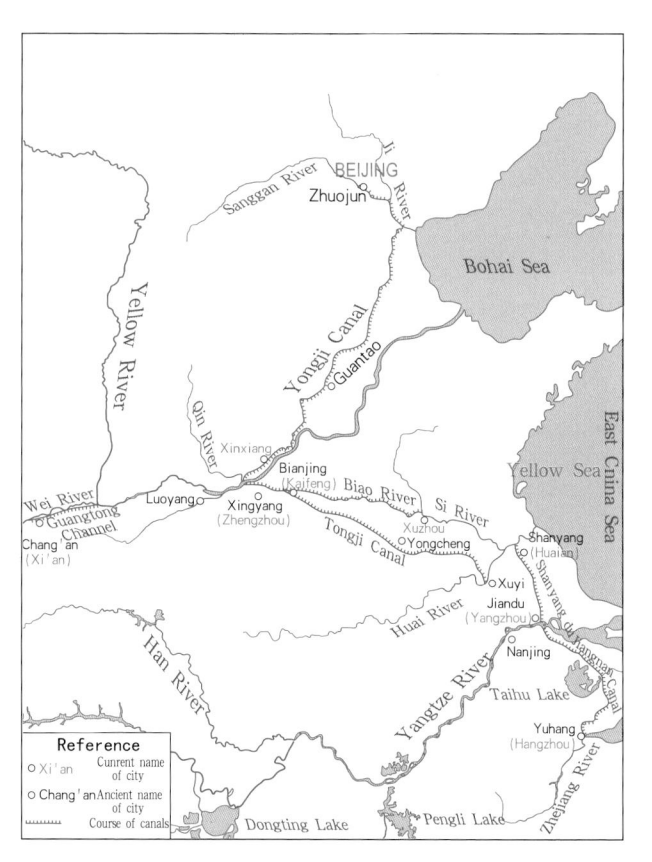

Fig.2　Map of the Grand Canal in the Sui, Tang and Song Dynasties (from the 6th to the 12th century)

the Yangtze and the Huai rivers, the Huai and the Yellow rivers, and the Yellow and Hai rivers respectively （Fig.1）. Based on these connections, the royal court in the Sui Dynasty (581 A.D. to 907 A.D.) ordered the excavation of the Tongji Canal and Yongji Canal and the rehabilitation of the Huaiyang Canal and the Jiangnan Canal, thereby forming a national Grand Canal system, with twin capital cities of Luoyang and Kaifeng at its center （Fig.2）. This enabled the shipment of food grain

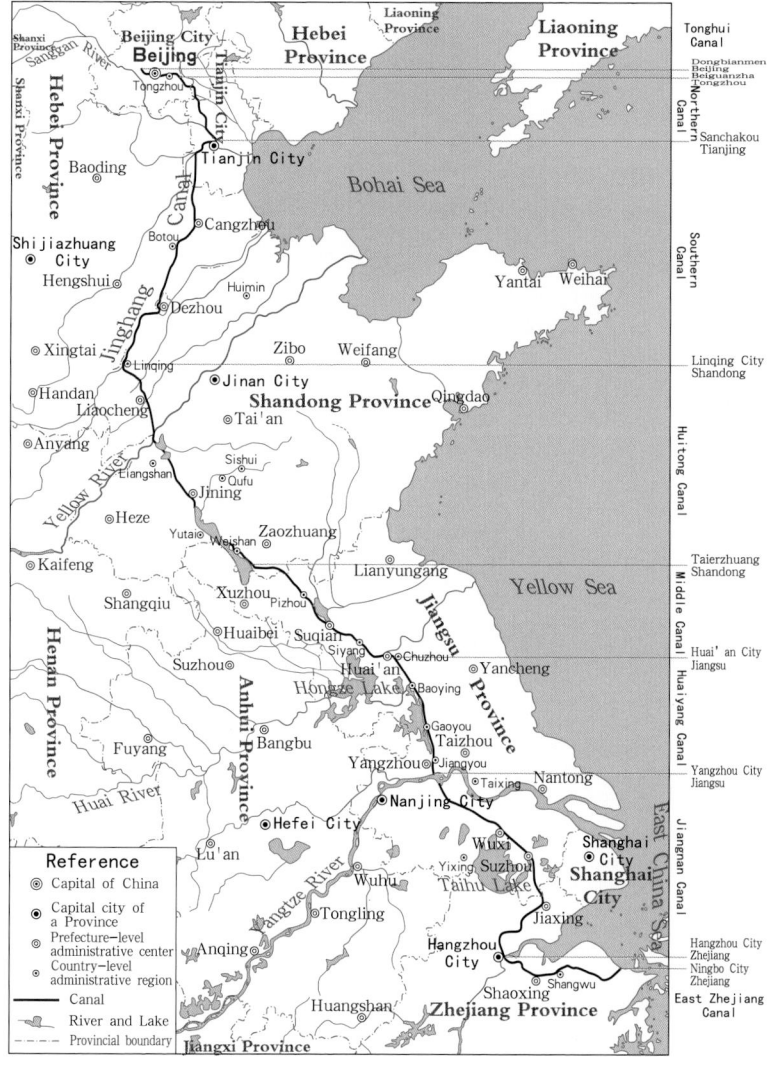

Fig.3　Map of the Grand Canal in the Yuan, Ming and Qing Dynasties (from the 12th to the 19th century)

from agricultural areas in the south to the northern part of the country. In Yuan Dynasty (1271 A.D. to 1368 A.D.), China's capital city was moved to Beijing. During this period, the curved route of Grand Canal formed in Sui Dynasty was straightened, and two new canals, the Huitong and Tonghui (Fig.3) were excavated and added to the system. As a result, the famous Beijing-Hangzhou Grand Canal system came into existence, and it remains well known to this day.

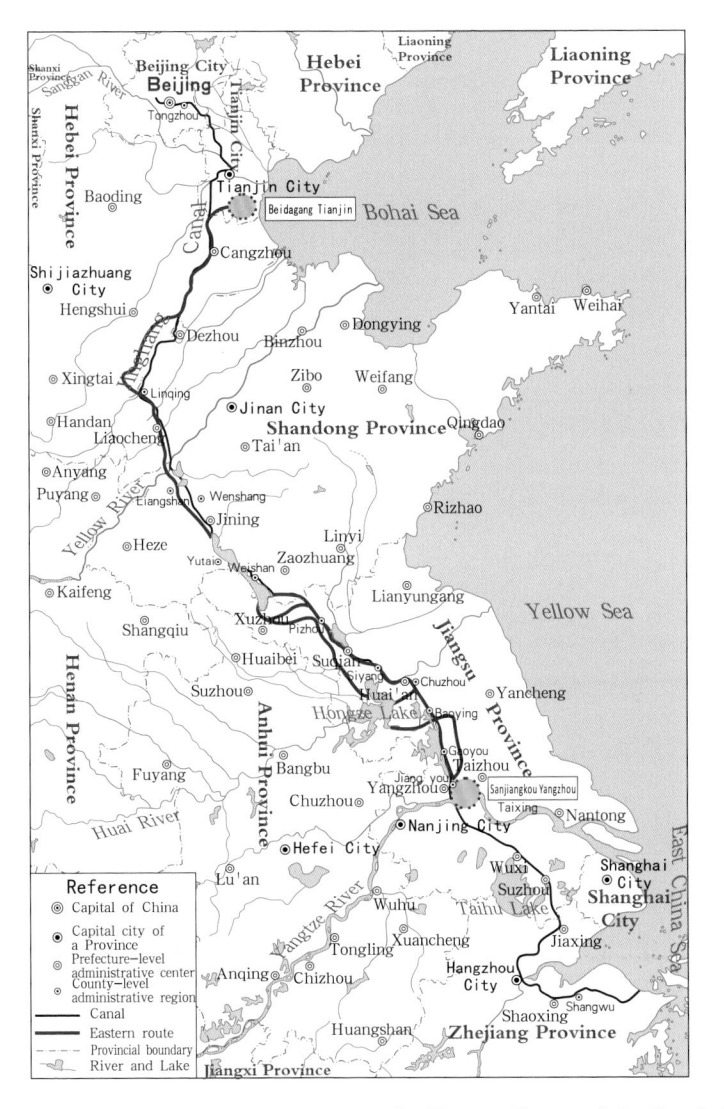

Fig.4　Comparison Map between the Eastern Route of the South-to-North Water Diversion Project and the Grand Canal

In order to facilitate the smooth functioning of the canal system, various water structures featuring regional characteristics, including sluices, dams and reservoirs, were built throughout different dynasties. These projects represent the cream of water science and technology innovation in Chinese history (Fig. 4 and Fig.5). Before the 1900s, the Grand Canal not only served grain transportation, but also boosted the cultural exchanges between southern and northern China. Being an economic and cultural nexus, it has made important contributions to the nation's social and economic development (Fig.6) .

Fig.5　The Relics of Pingjin Sluice Gate of Tonghui River

Fig.6　Picture Depicting Ships Sail Through the Dam, Painted by British Mission to China During Qian Long's Rule in Qing Dynasty

Four

Qiantang River Seawalls — Dykes with 2,000 years of history

Qiantang River, the largest river in China's Zhejiang Province, is named after the ancient county of Qiantang (today's city of Hangzhou). With a catchment area of 55,058 km^2, the Qiantang River runs for 588.73 kilometers from its northern source to its estuary at the Xin'an River. It flows through Anhui and Zhejiang provinces, and finally empties into the East China Sea at Hangzhou Bay.

The Seawalls of the Qiantang River are dykes built on both sides of the Qiantang Estuary to resist strong tidal erosion. As one of the natural wonders of the world, the river tide of the Qiantang River is known as

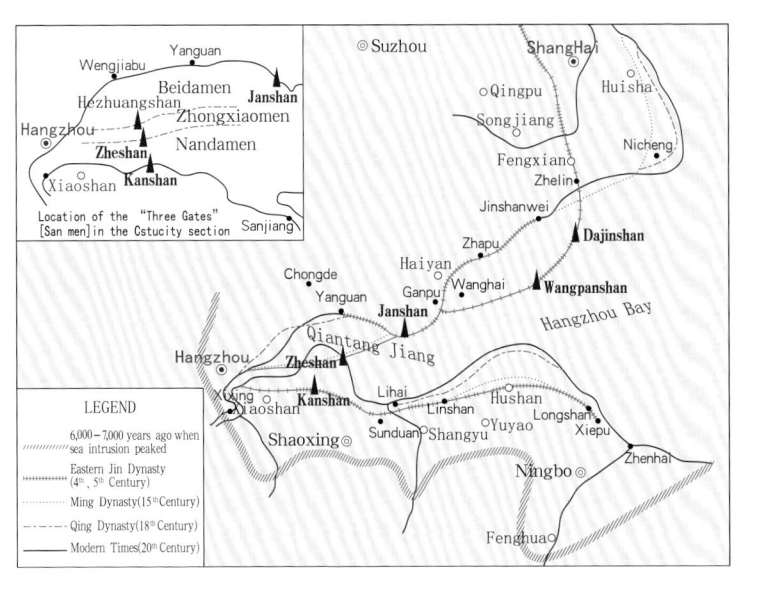

Fig.1   Evolution of the Shoreline at the Qiantang River Estuary and the Location of the "Three Gates" [San Men](from Haining City Administration of Cultural Heritage)

"the champion tide on earth". It is formed by tidal currents entering the estuary under the centrifugal effect of gravity and the rotation of the Earth, in addition to the special topographical features of the estuary. On the occasion of the astronomical tides, the tidal bore can reach as high as nearly 10 meters, which would destroy nearby homes and farmland if no seawalls had been constructed. Seawalls were built and reinforced during each and every dynasty in order to withstand the intrusion of salt tides.

The Seawalls of the Qiantang River, which are mostly earthworks, were first built in the Eastern Han Dynasty (25 A.D. to 220 A.D.) and gradually expanded along both banks of the river. Stone seawalls appeared during the Tang Dynasty (618 A.D. to 907 A.D.), In the fourth

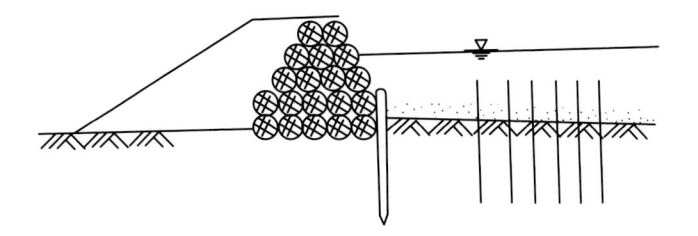

Fig.2 "Bamboo Cage Stone Wall" Invented by Qian Liu, King of Wuyue

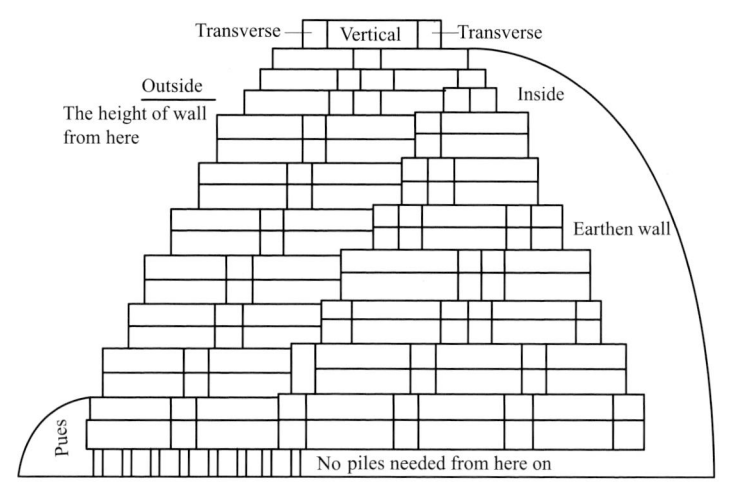

Fig.3 The Ming Dynasty Defense-in-depth System Against Tides of the Qiantang River

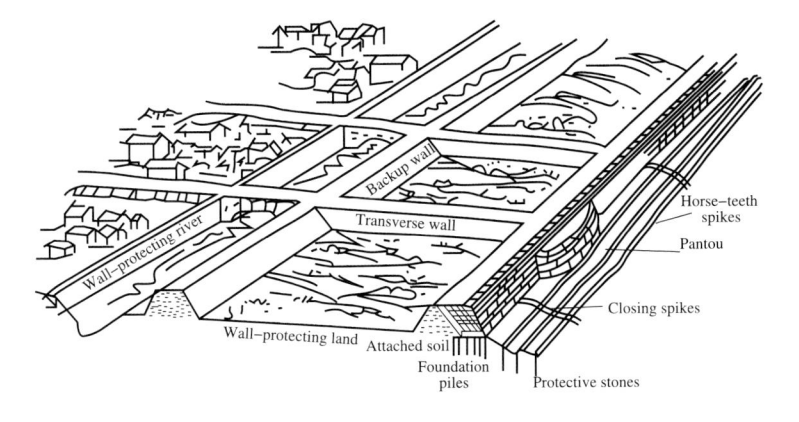

Fig.4　Five Vertical and Five Horizontal Fish-scale Stone Wall Designed by Huang Guangsheng in the Ming Dynasty

year of the reign of Emperor Kaiping (910 A.D.) during the Liang Dynasty (907 A.D. to 923 A.D.), Qian Liu, ruler of the small kingdom of Wuyue, invented "Bamboo Cage Stone Wall" (Fig.2). During the Northern Song Dynasty (960 A.D. to 1127 A.D.), firewood, willow, vertical stone and slope stone walls were subsequently adopted. Five vertical and five horizontal fish-scale stone walls (Fig.3) prevailed during the Ming Dynasty (1368 A.D. to 1644 A.D.). In the Qing Dynasty (1616 A.D. to 1912 A.D.), the local government developed a tidal wave defense-in-depth system (see Fig.4) consisting of a foundation project, main wall, backup wall, traverse wall, backup wall river and wall protection works.

After the establishment of General Administration for Seawalls of Zhejiang in the 34th year of Emperor Guangxu's reign (1908 A.D.) in the Qing Dynasty, new foreign technologies were introduced for the construction of concrete gravity walls (Fig.5) in Haining County from 1919 to 1927. After the founding of the People's Republic of China, the seawalls were completely renovated, thanks to the repairing and strengthening of old, scattered and branch dykes. The construction of standardized walls began at the end of the 20th century (Fig.6 and Fig.7).

Fig.5   The Tide of Qiantang River

Fig.6   The Great Fish-scale Seawall

Fig.7   Xiaoshao Seawall

The maintenance of the Qiantang Seawalls has always been a governmental priority throughout the history of China. Since the Tang Dynasty, the management of seawalls has been recorded. Regular repair and maintenance was carried out by the county government from the Song Dynasty (960 A.D. to 1279 A.D.) to the Ming Dynasty. The seawalls were coded section by section in Haiyan County during the Ming Dynasty, each section measuring 20 zhang (the equivalent of 3.33 meters), in order to clarify the specific location of each section and exercise section-based management. A dedicated commissioner was appointed by the central government to manage the seawalls during the Qing Dynasty. After the founding of the People's Republic of China, the Qiantang River Authority was established. In order to ensure unified management of the seawalls, mileage pegs were used to replace the thousand-character monuments.

Five

The Three Gorges Project — The pillar of China

Located in the upper mainstream of the Yangtze River, the Three Gorges Project passed its final check and acceptance in September 2015. Since the opening of the Three Gorges Reservoir in June 2003, it has been operating safely for 16 years. Over this period, it has contributed greatly to flood control by defending the area from a flood in 2010 which was the worst in two decades, as well as two major floods in 2018. At the same time, it has generated over 100 billion kWh of green electricity.

People may wonder why such a project is named the Three Gorges. The name refers to the three splendid valleys of the Yangtze River where the project is situated. From west to east, they are Qutang Gorge, Wu Gorge and Xiling Gorge, which people collectively refer to as the "Three Gorges". Now, the Three Gorges Project adds new elements to the beauty of the Yangtze River Basin, and Chairman Mao Zedong once envisaged that "a project may break the wind and rain around Wushan Mountain, thus a plain lake surface will appear high above the valleys".

The construction of a water project on the mainstream of the Yangtze River needs not only bold ideas, but also scientific foundations. This bold idea was first put forward by Dr. Sun Yat-sen (the pioneer of the great democratic revolution in modern China) in his 1918 book entitled "A Plan for Nation Building". Before 1949, the government of the Republic of China began to carry out feasibility studies for

Fig.1   The Three Gorges Project

constructing such a project. Moreover, systematic scientific studies had been practiced from 1986 to 1992, and Chinese scientists conducted comprehensive and repeated rounds of feasibility studies on 14 subjects including biology, geology, sedimentation and water quality. Finally, project scale and main functions in relation to flood control, power generation, navigation and water resource utilization were determined.

The Three Gorges Project is mainly composed of three major parts: the dam project, the resettlement project, and the power transmission and transformation project. The dam crest elevation is 185 meters, the water storage elevation is 175 meters, the dam length is 2,309.5 meters, the total installed capacity is 22.4 million kW, the total storage capacity is 39.3 billion cubic meters, the maximum navigable tonnage is 3,000 tons, and the static investment amounts to 135.266 billion RMB. After the completion of the Three Gorges Project, the flood control standard of Jingjiang section of the Yangtze River was raised to once in every 100 years. Its annual average capacity of power generation is 84.88 billion kWh, equivalent to reducing annual coal consumption by 50 million tons.

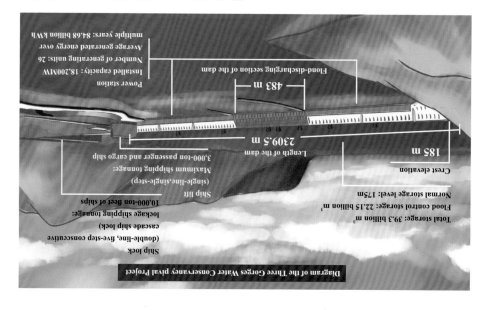

Diagram of the Three Gorges Water Conservancy pivral Project

Ship lock
(double-line, five-step consecutive
cascade ship lock)
lockage shipping tonnage:
10,000-ton fleet of ships

Ship lift
(single-line,single-step)
Maximum shipping tonnage:
3,000-ton passenger and cargo ship

Power station
Installed capacity: 18,200MW
Number of generating units: 26
Average generated energy over
multiple years: 84.68 billion kWh

Flond-discharging section of the dam

483 m

Length of the dam
2309.5 m

Crest elevation
185 m

Normal storage level: 175M
Flood control storage: 22.15 billion m³

Total storage: 39.3 billion m³

Fig.2   Diagram of the Three Gorges Project

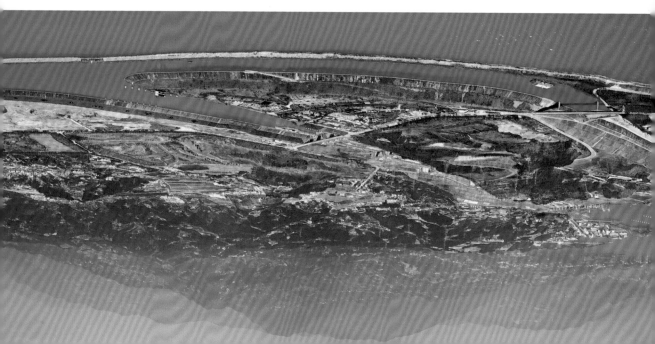

Six

Lingqu Canal — The world's most complete ancient man-made canal

Lingqu Canal, located in Xing'an County, Guangxi Zhuang Autonomous Region, is one of the most completely preserved ancient man-made canals in the world today. It has been acclaimed as one of the Three Major Water Projects of the Qin Dynasty (221 B.C. to 207 B.C.) in China, together with the ancient Dujiangyan Project and the Zhengguo Canal. In 2018, it was added to the World Irrigation Project Heritage List.

In 219 B.C., the Emperor of the Qin Dynasty in China started to build

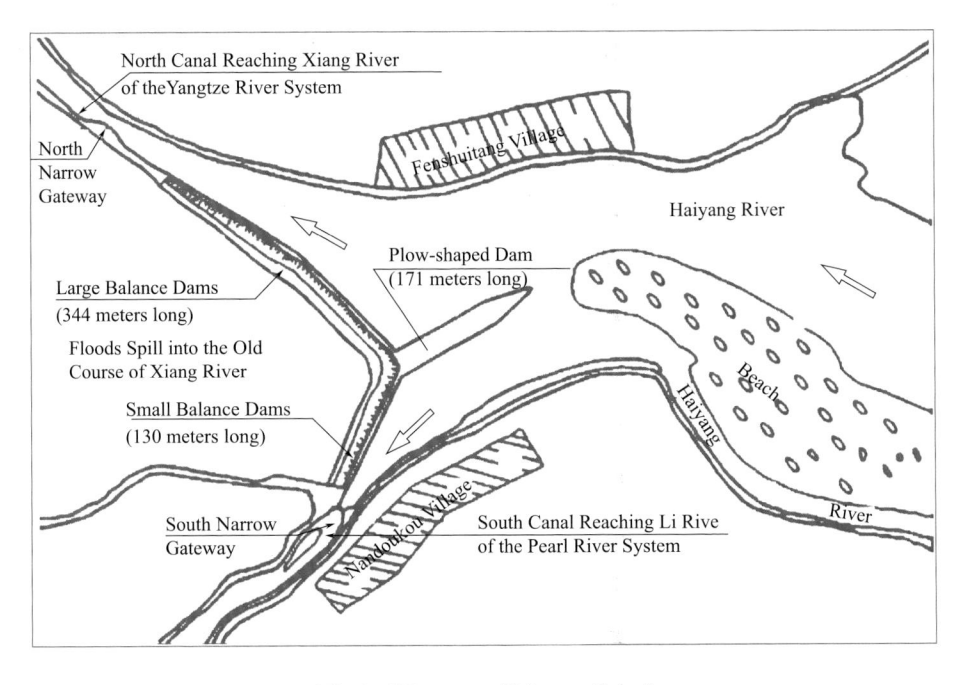

Fig.1　Diagram of Lingqu Canal

the Lingqu Canal, in order to unify the southern region and facilitate the transportation of food grain to supply the army. It linked the Haiyang River – the source of the Xiang River (the largest tributary of the Yangtze River) and the Darong River – the source of the Li River (the fifth largest tributary of the Pearl River). The Lingqu Canal had a total length of 36.4 kilometers, and opened to navigation in 214 B.C. after five years of construction. Upon completion of the canal, the Yangtze River basin and the Pearl River basin were connected, thus forming a navigation network across eastern and southern China.

The main part of the Lingqu Canal includes the canal head, the south canal and the north canal. Its guiding design principle is to intercept the Xiang River with a water diversion structure consisting of a rafter dam (large and small balance dams) and a plow-shaped dam, which could divert 30% of the river runoff into the Li River through the south canal (Fig.1).

The selection of the site of the Lingqu Canal, its ingenious design and precise construction of demonstrate the oustanding engineering of ancient water projects in China. The project layout and hydraulic

Fig.2　Lingqu Canal

structures particularly reflect the unique cultural perspective of the Chinese people in terms of water project design, technological sophistication and aesthetic values. It epitomizes the marvelous achievements in the spheres of navigation and hydraulic engineering in ancient China. The steep sluice gate, which was built to raise the

Fig.3　The Stone Stele Marked the Diversion Between the Xiangjiang River and the Lijiang River

water level and ensure navigation, was recognized as "the Father of the Sluice Gate" by experts of the International Commission on Large Dams.

Although navigation no longer takes place on the Lingqu Canal, it continues to perform its irrigation function and contributes to the welfare of local people.

Fig.4　Lingqu Canal Head

Seven |

Dongfeng Weir
— A World Irrigation Heritage Project

Dongfeng Weir, located in Jiajiang County of Sichuan Province, is a multipurpose project, which has been used for more than 350 years, mainly serves irrigation and other flood control purposes. In 2014, thanks to its long history, perfect canal system and remarkable comprehensive benefits, it was added to the World Irrigation Project Heritage List.

Fig.1　Dongfeng Weir

The Jia River is located in the transition section between the Qinghai-Tibet Plateau and Sichuan Basin, and hydrologically in the transition section between the midstream gorge of the Qingyi River and its downstream flatland. After the Qingyi River, a tributary of the Yangtze River, runs through its last gorge of the Qianfoyan in Jiajiang, its river course widens and its flow slows down, providing fertile land and abundant water sources for people in the area. The favorable climate and the appropriate water head drop enabled local people to divert water by gravity for irrigation purposes.

Historically, local people had emulated the Dujiangyan Project by filling bamboo cages with rocks to dam water along the Qingyi River, digging various water intakes, and building a crisscrossing network of large and small canals. As it did in the Chengdu Plain area, this also transformed Jiajiang County into a land of abundance "free from floods, droughts and hunger". Nevertheless, most of the ancient weirs

Fig.2　Inscriptions about Water Engineering

in Jiajiang were excavated by the local people and not managed in a unified manner, resulting in poor management and operation. People struggled to get more water during the peak season for use, while fighting among themselves during the flood season and even benefit themselves at others' expense. Water-induced disputes emerged one after another. The key to solving these problems was to have an integrated plan, the implementation of which required not only organization by the

Fig.3   The Trunk Canal of Dongfeng Weir and Qianfo Cliff

government, but also certain technologies. During the Zhengtong Period of the Ming Dynasty (1436 A.D. to 1449 A.D.), Lu Lun, the county magistrate, became the first person to take this key step which marked the origins of the construction of the Dongfeng Weir.

By taking the advantages of natural beaches, Lu Lun led local people to store stones in bamboo cages in an orderly pattern and divert water from the Chahe River on the left bank of the Qingyi River without requiring the construction of dams. For the first time in the history of Jiajiang, the numerous weirs and canals that people had excavated spontaneously since ancient times were classified into two large canals, which changed the decentralized management into unified management. The two weirs excavated by Lu Lun form the major components of the Dongfeng Weir, namely, three big weirs and eight small weirs. The three big weirs, Shijie Weir, Yongtong Weir and Longxing Weir, constitute

a trunk canal to irrigate the southern part of the county, and each weir irrigates a township, while eight small weirs form a trunk canal that irrigates the east and north of the county.

Around the turn of the Ming (1368 A.D. to 1644 A.D.) and Qing (1644 A.D. to 1911 A.D.) dynasties, Sichuan suffered from intense wars, and the irrigation canal system in Jiajiang County was seriously damaged. In 1662 (the first year of the reign of Emperor Kangxi in the Qing Dynasty), Wang Shikui was the first magistrate of Jiajiang County to have been dispatched by the government of the Qing Dynasty. Wang reconstructed three big weirs that had been destroyed in wars. Bamboo cages were filled with pebbles and built into a 300 m-plus diversion weir at the intake of the upstream tributary of the Qingyi River. The backwater was diverted into trunk canals of three big weirs and eight small weirs. The southeast main weir is named Pilu Weir because it is near the Pilu Temple. Pilu Weir is the prototype of the Dongfeng Weir. Therefore, 1662 is regarded as the official start of the excavation of the Dongfeng Weir.

However, water diversion from Pilu Weir was unable to keep up with growing needs from irrigation. In order to ensure the enduring use of this irrigation system, the Liugong Weir was added, and three big weirs and eight small weirs were continuously innovated. The water intake was also moved upwards. In the 1970s, this irrigation system was renamed the Dongfeng Weir.

Eight

Zhengguo Canal — A large dam-free water diversion irrigation project with a history of over 2,200 years

Construction of the Zhengguo Canal, a large dam-free water diversion irrigation project, began in 246 B.C. It played an important role in the prosperity of the Qin Kingdom (770 B.C. to 221 B.C.) and the unification of China under the Qin Dynasty (221 B.C. to 207 B.C.) .

Located in the central plain of present-day Shaanxi Province, the Zhengguo Canal, which is around 150 km long, diverts water from the Jing River in the west to the Luo River in the east. Since its completion more than 2,200 years ago, a large number of canals were built, namely Bai Canal, Zhengbai Canal, Fengli Canal, Wangyushi Canal, Guanghui Canal and Jinghui Canal, and most of these continue to function for irrigation purposes.

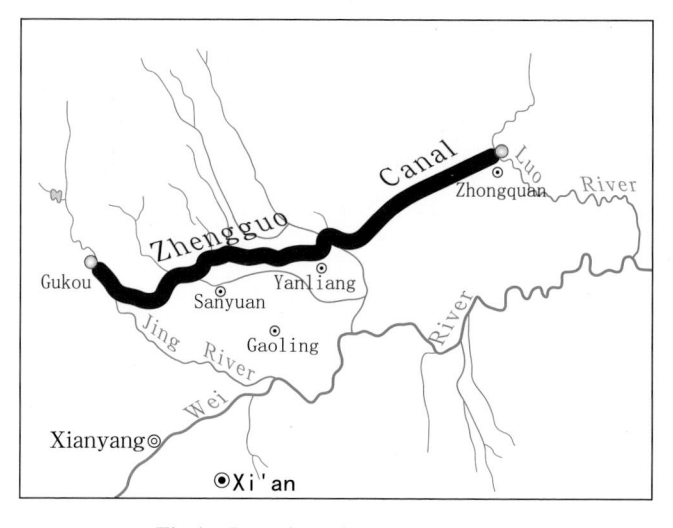

Fig.1　Location of Zhengguo Canal

The Jing River, which is the source of the Zhengguo Canal, is well-known for its high concentration of sediment, with an annual sediment discharge of 265 million tons. In ancient time, people already knew that river sediment had rich organic matters and could be silted to fertilize lands. Thanks to the construction of the Zhengguo Canal, water with high sediment content from the Jing River could be diverted for the reclamation of saline-alkali land by silting irrigation. Due to the benefits brought to the area, the Zhengguo Canal was praised for irrigating and fertilizing fields, nurturing the rapid growth of crops, feeding and

Fig.2　Zhengguo Canal

clothing the people, and ensuring the growth of the population.

Zhengguo Canal was once an irrigation scheme managed by ancient kingdoms and dynasties. Water laws and regulations, including Water Department Regulation and Six Laws, unveiled by the central government of the Tang Dynasty, made special provisions for the management of the Zhengguo Canal, including the allocation of irrigated water. These prioritized irrigation water over the operation of water mills, which showed a greater awareness of water rights.

The earliest diversion outlet of the Zhengguo Canal is situated in Wangqiao Town, Jingyang County, where the Jinghe River run out of gorges. After more than 200 years, the diversion outlet moved upwards 1.3 kilometers, which is known as the Bai Canal. After over 800 years, the canal head moved up to the gorge section of the Jing River in the 11th century. In the 14th century, it shifted upwards another 990 meters, together with the excavation of tunnels, to deliver water. In 1932, the irrigation canal diverting water from Jing River was rebuilt. Over the past 2,000 years, the riverbed of the Jing River has been deepened by around 20 meters, and the canal head has moved upwards about 5 kilometers compared with its original location.

Nine

Quebei — A water storage system with 2,600 years of history

Quebei, also known as Anfeng Pond, is a reservoir and irrigation system located in the Huaihe River Basin, which 30 kilometers south of Shou County in Anhui Province. Since its construction began in 598 B.C., Quebei has been used for irrigation and it remains operational to this day. Quebei, which is China's oldest reservoir as it is 300 years older than the Dujiang Weir and Zhengguo Canal, currently irrigates more than 670,000 mu (44,700 ha). It was added to the World Heritage Irrigation Project List in 2015.

In the 6th century B.C., Sun shu'ao, prime minister of the Chu State in the Eastern Zhou Dynasty, built the reservoir for water storage and

Fig.1　Quebei 1

irrigation by taking the advantage of low-lying land. Quebei had a large water surface in the period soon after its completion, but it was gradually reclaimed as farmland due to the demand for food and local population growth. By the 6th century, the water storage area had been halved; by the end of the 16th century, it was just over 40 square kilometers, and it is only 34 square kilometers today.

The maximum storage capacity of

Quebei reached 90.7 million cubic meters. In ancient times, its water source partly came from southern mountain streams, from the Shanyuan River that originated in Dabie Mountain, and partly diverted water to the Piyuan River. Since Quebei was incorporated in Pi-Shi-Hang Irrigation District in 1958, the Pi River has become its main source and entered through the main channel of the Pidong to the Quebei system, with the water mostly being transported by gravity.

The Quebei irrigation system mainly consists of a storage reservoir, irrigation gate, irrigation and drainage canal system and flood control projects. Surrounded by levees on its northern, eastern and western sides, it has a total length of 26 kilometers and its form and structure are almost the same as in the 19th century.

Quebei is a large-scale infrastructure serving public good. It was mainly constructed by the central government, with operation and

Fig.2   Quebei 2

maintenance of the irrigation system involving cooperation between government and the private sector. Across different dynasties, governments at all levels have been in charge of the construction, operation and maintenance of reservoirs, irrigation gates and main canals, and making regulations. In the 2nd century B.C. during the reign of Emperor Wu of the Han Dynasty, an official was designated in particularly to manage Quebei. In the 1950s, an iron tally of the Eastern Han official for irrigation management was discovered at the Quebei Levee, which testified to the authoritativeness of the government at that time.

The famous Chinese hydraulic engineer Wang Jing, who had been in charge of the management of Quebei, established rules for its annual maintenance and inscribed them on a stone tablet as a public notice. A farmers' organization was set up in the irrigation district to manage and maintain the order of water use. The partnership between government and the private sector continue to function with the Anfengtang division, a government body, in charge of the maintenance of storage works, and the main and branch canals; while people benefiting from the project are responsible for managing canals below the branch level and water distribution systems.

Construction of Quebei made the State of Chu prosperous 2,600 years ago. About 600,000 people in 13 townships and 114 villages spread over the Quebei Irrigation District currently benefit from the project.

Ten |

Xiaolangdi Hydraulic Complex — The China's largest rockfill dam

As one of the largest hydraulic complexes located on the middle reaches of the Yellow River, the Xiaolangdi Hydraulic Complex has played a significant role in flood protection and water-sand regulation for the downstream areas of the basin. Situated at the last canyon of the Yellow River, it is only 40 km away from the north of Luoyang, one of China ancient capitals with more than 1,000 years of history.

The Yellow River is known for its heavy sedimentation, which has resulted in a hanging river that threatens people's lives and property in

Fig.1 Xiaolangdi Multi-Purpose Water Project

the downstream. Since the dam was completed in 1999, the Xiaolangdi reservoir has been used to regulate water and sand. By adjusting the sluice gates, operating water discharge and flow rate, soil and sand deposited in the reservoir have been discharged, in order to slow the further deposition of sand in the downstream river bed.

With a total capacity of 50.73 million m^3, Xiaolangdi is China's largest rockfill dam. The complex comprises of a large number of gates for the discharge of water and sediment, as well as a hydropower plant with 6 units and a total installed capacity of 1,800 MW.

Due to its unfavorable topographical and geological conditions, as well as the complicated correlation between water and sediment, the construction of Xiaolangdi is considered the most challenging in the world.

The Xiaolangdi complex has generated great social and economic benefits since becoming operational. More than 300 million tons of sediment have been washed away from the riverbed in the downstream, thanks to water-sediment regulation. It not only plays a crucial role in flood control and drought relief, but is also a key element in ensuring ecological security.

Eleven |

Mulanbei — An ancient water project which continues to serve modern society

Construction of Mulanbei started in 1064, with the objective of blocking twice-a-day salt tides and retaining precious fresh water resources. The floodwater may run through the Mulan Sluice whenever it occurs.

Mulanbei in central Fujian Province, is named after the Mulan River which flows into the sea without joining other rivers. Mulanbei was constructed to make full use of the river water for irrigation purposes. It supplies water to 163,000 mu of farmland and for domestic use to more than 800,000 people in 12 towns.

The amazing aspect of Mulanbei is reflected by its appropriate dam height. The gravity dam and its foundation were built with local granite blocks, to enable the gravity hydraulic structure to withstand the impact of high-speed floods. The wooden and riprap-filled foundation of the dam ensure that no significant sinking of large pieces of stonework occur, after the dike and stonework were anchored with iron. Although it is about 900 years old, the Mulan Dam still appears as it was originally

Fig.1 The Status of Qian Siniang, the Original Founder of Mulanbei

built.

The original Mulanbei was constructed by a woman named Qian Si-niang, but it was destroyed by floods. Its reconstruction in its present position was initiated at the start of the Xining era of the Song dynasty (1075 A.D.) by a Buddhist monk with a knowledge of hydraulic engineering. After 8 years, Mulanbei was reconstructed in 1083.

Fig.2　The Diversion Dike of the Mulanbei

Twelve |

Water runs outside the polders and farmland is developed inside the polders—Lougang Project in Taihu Lake

Taihu Lake is situated between Wuxi and Suzhou in Jiangsu Province and Huzhou in Zhejiang Province. As early as the Western Jin Dynasty (265 A.D. to 316 A.D.), its water area was larger than it is now. A large amount of migrants rushed to the Taihu Lake basin and found that there was more water and less land, especially the vast shallow water area. How could they separate the soil from the water and use this part of land? Our ancestors explored an ingenious way, which is what people call Lougang today.

They built two permeable walls made of bamboo and wood. The soft liquid soil in the middle was dug outside the retaining walls, while water in the mud penetrated through the bamboo and wood wall into the channel of two walls, and then flowed into the lake. In this way, such retaining walls were made on the beach in rows around Taihu Lake.

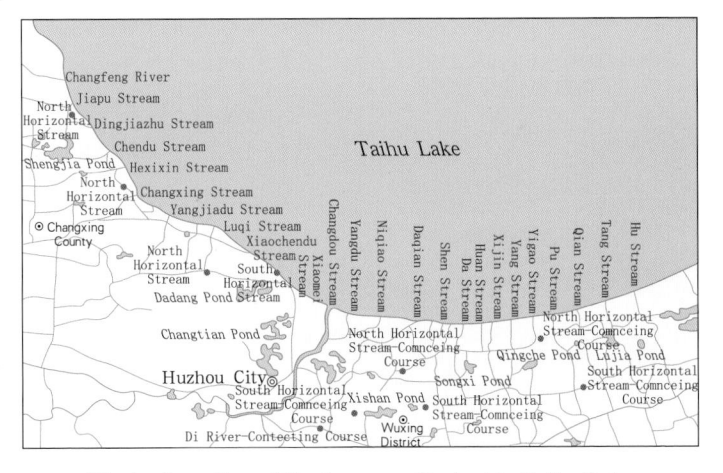

Fig.1　Location of the Lougang Project in Taihu Lake

The land around Taihu Lake was subsequently divided block by block. People called such walls "polders". Water runs outside the polders, and farmland is developed inside the polders.

The watercourse between the two walls is called Lougang. In order to ensure its safety, people built sluices at the junction of Lougang and Taihu Lake. The opening and closing of the sluice can adjust the water level of Lougang. The sluice was opened to bring flood water into the lake in case of floods in the upstream of Lougang, while the lake water would be diverted into Lougang for irrigation when its water level is low.

Taking a bird's eye view of Lougang on the south bank of Taihu Lake, it can be seen that all the directions of Lougang entering the lake are from the northeast. This is because the northwest wind prevails in the south bank area in the winter. People use it to wash sediment from the south into the lake to prevent sediment deposition in Lougang. In addition, Lougang mostly becomes narrow at the entrance to Taihu Lake, as the suddenly cramped watercourse will speed up the flow and flush the mud away, thus reducing the cost of dredging. Lougang is now regarded as an ancient drainage and irrigation project and has been added to the World Irrigation Project Heritage List in 2016.

Fig.2　The Lougang Project in Taihu Lake

Thirteen

Ningxia Ancient Yellow River Irrigation Area — Water culture through thousands of years

The ancient Yellow River irrigation area in Ningxia is one of the four ancient irrigation sites of China. Consistent with the terrain of the Yellow River, it covers a J-shaped strip between Xiaheyan and Shizuishan, two hydrographic stations in the upper reach of the Yellow River, which have irrigated the area for more than 2,000 years. The topography of the Ningxia Plain is higher in the southern part than in the north, thus forming a sloping landform. Taking advantage of the unique geographical conditions, the smart ancient people diverted water from the natural moat into canals, and then the land became fertile due to the exclusive advantages of abundant irrigation, creating a miracle in which "the Yellow River only makes Ningxia rich". That's why people often say "the Yellow River does much harm, but makes the Plain rich".

In early years, the ancient Yellow River irrigation area covered only the southern part of the Yinchuan Plain. Afterwards, the people of many generations and dynasties dredged old canals and dug new ones to continuously expand the irrigation coverage. Up to the prosperous Tang Dynasty in the 7th century, a total of 13 trunk canals were built, irrigating an area of 1 million mu (1 ha equals 15 mu), and a gravity irrigation system was preliminarily formed in the Yinchuan Plain and the Weining Plain. During the Western Xia Dynasty in the 11th century, its engineering and management systems were further improved, and the irrigation area increased to 1.6 million mu. During the Yuan, Ming

and Qing dynasties, the number of trunk and branch canals continued to increase. At the end of the 19th century, more than 20 trunk canals existed, with a total length of more than 1,500 km and a total gravity irrigation area of 2.1 million mu.

The ancient Yellow River irrigation area consists of a canal system, drainage system, gates and other regulating facilities. The ancient Yellow River irrigation mainly features diversion without dams. Starting from the head of each canal, people used stone to build an embankment which was several kilometers long in order to split the river and divert water, which is known as the "water-diverting weir", and a spillway was built on the weir to guarantee the safety of the canal mouth in case

of flooding.

After the Qingtongxia and Shapotou hydraulic engineering projects were built up in 1968 and in 2004 respectively, all the diversionwithout-dam projects were replaced by diversion-with-dam ones, and all the canal projects adopted controlling gates, expanding the irrigation area and improving the irrigation dependability through system transformation. The irrigation area currently covers 12,953 km^2 and irrigates 8.28 million mu. It consists of 25 trunk canals, with a total length of 2,454 km, a total water diversion capacity of 750 m^3/s, and 9,265 control facilities. Today, the Ningxia irrigation site is one of the 12 commodity grain bases in China.

Fig.1　The Ancient Yellow River Irrigation Area in Ningxia